THE POETRY OF IODINE

The Poetry of Iodine

Walter the Educator

SKB

Silent King Books a WhichHead Imprint

Copyright © 2023 by Walter the Educator

All rights reserved. No part of this book may be reproduced in any manner whatsoever without written permission except in the case of brief quotations embodied in critical articles and reviews.

First Printing, 2023

Disclaimer
This book is a literary work; poems are not about specific persons, locations, situations, and/or circumstances unless mentioned in a historical context. This book is for entertainment and informational purposes only. The author and publisher offer this information without warranties expressed or implied. No matter the grounds, neither the author nor the publisher will be accountable for any losses, injuries, or other damages caused by the reader's use of this book. The use of this book acknowledges an understanding and acceptance of this disclaimer.

"Earning a degree in chemistry changed my life!"
- Walter the Educator

dedicated to all the chemistry lovers, like myself, across the world

CONTENTS

Dedication V

Why I Created This Book? 1

One - Iodine's Power 2

Two - Revealing The Secrets 4

Three - Reign With Beauty 6

Four - Waiting To Be Told 8

Five - Forever We Chase 10

Six - Wonders Of Life 12

Seven - Time And Space 14

Eight - Magic 16

Nine - Companion To Science 18

Ten - Sway 20

Eleven - Forever In Our Heart 22

Twelve - Symbol Of Possibilities 24

Thirteen - Sweetest Wine	. . . .	26
Fourteen - Beacon Of Creativity	. . . . .	28
Fifteen - Always Be There	. . . .	30
Sixteen - Silent Force	. . . .	32
Seventeen - Enchanting The World	. . . . .	34
Eighteen - Its Presence	. . . .	36
Nineteen - Iodine, The Element	. . . .	38
Twenty - Hush	. . . .	40
Twenty-One - Wonders Of Iodine	. . . . . .	42
Twenty-Two - Chemical Marvel	. . . .	44
Twenty-Three - Muse Supreme	. . . .	46
Twenty-Four - Enthrall	. . . .	48
Twenty-Five - It Shall Be	. . . .	50
Twenty-Six - Wonder And Grace	. . . . . .	52
Twenty-Seven - Will Stand	. . . .	54
Twenty-Eight - Science And Dream	. . . . .	56
Twenty-Nine - Creativity's Realm	. . . . . .	58
Thirty - Iodine, The Element That Ignites	. .	60
Thirty-One - Leave Us In Awe	. . . .	62
Thirty-Two - Spark	. . . .	64

Thirty-Three - Unfold 66

Thirty-Four - Essence Endures 68

Thirty-Five - Shining Ever Bright 70

About The Author 72

WHY I CREATED THIS BOOK?

Creating a poetry book about the chemical element of Iodine was unique and intriguing. Iodine, as a chemical element, possesses its own characteristics, symbolism, and historical significance. By exploring these aspects through poetry, I can weave together a compelling narrative that combines science, emotions, and art. Additionally, delving into the properties, uses, and effects of iodine can offer a rich source of inspiration, allowing for exploration of themes such as transformation, healing, and the interplay between light and darkness. Ultimately, this poetry book can offer readers a fresh and insightful perspective on both science and the human experience.

ONE

IODINE'S POWER

In the depths of the ocean, a secret dwells,
A shimmering element, where the mystery excels.
Its name is Iodine, noble and bold,
A treasure untold, a story yet to be told.

A lustrous hue, like the morning sun,
Iodine's brilliance, second to none.
In nature's embrace, it finds its way,
A vital ingredient, where life holds sway.

From the sea's embrace, it rises high,
Carried on the wind, across the sky.
It whispers of healing, of strength and might,
Iodine's essence, a beacon of light.

In the human body, a vital role it plays,
Regulating functions, in countless ways.
Thyroid, the guardian, it guards with care,
Ensuring balance, beyond compare.

In laboratories, it finds its home,
A marvel of science, where discoveries roam.
Revealing secrets, unlocking the unknown,
Iodine's magic, forever shown.

So let us celebrate this element true,
With reverence and awe, in all we do.
For Iodine's power, both gentle and grand,
Unveils the wonders of nature's hand.

TWO

REVEALING THE SECRETS

In the realm where nature weaves its tale,
There lies a mystic, Iodine, so pale.
A shimmering element, with power untold,
Unveiling secrets, mysteries unfold.

    Within thy depths, the ocean's embrace,
Iodine resides, with ethereal grace.
A treasure hidden, beneath the waves,
A testament to nature's enigmatic ways.

    In vibrant forests, where life abounds,
Iodine dances, in the leafy grounds.
From the soil to the sky, its presence clear,
A catalyst for growth, both far and near.

    Within our bodies, a sacred bond,
Iodine's magic, forever beyond.

Thy presence vital, for thyroid's might,
A symphony of balance, day and night.
    Scientific minds, with keen insight,
Unveiled the wonders, Iodine's light.
From Mendeleev's table, it took its place,
A testament to science's boundless grace.
    So let us celebrate, this element divine,
Iodine, the marvel, let it forever shine.
In nature's tapestry, a jewel so rare,
Revealing the secrets, we all long to share.

THREE

REIGN WITH BEAUTY

In the depths of the ocean's embrace,
Where the waves dance and shadows chase,
Resides a jewel, a secret untold,
A mystic element, a tale to behold.
    Iodine, the enigmatic sprite,
Bathed in moonlight, a shimmering light,
A touch of magic, a whisper of grace,
It weaves its spell in nature's embrace.
    In the silent forests, where secrets lie,
Iodine paints the leaves with a vibrant dye,
A tapestry of hues, a symphony of green,
In this dance of life, Iodine is seen.
    A guardian of balance, a healer's balm,
Within the human body, it imparts its calm,

From thyroid's throne, it reigns supreme,
Regulating the rhythm of life's grand theme.
   In laboratories, where discoveries thrive,
Iodine's essence helps science to thrive,
A catalyst for change, a key to explore,
Unlocking mysteries, forevermore.
   Oh, Iodine, your mystery profound,
In your presence, wonders are found,
From the depths of the ocean to the human soul,
You reign with beauty, your story untold.

FOUR

WAITING TO BE TOLD

In nature's vast and wondrous realm,
There lies a treasure, Iodine, the helm.
A trace element, so small, yet grand,
With powers that few can understand.

Within the depths of oceans deep,
Iodine dwells, its secrets to keep.
It colors the waters with a mystical hue,
A gift from nature, ever true.

In forests dense, where shadows play,
Iodine whispers, night and day.
It weaves its magic through the trees,
Bringing balance, healing, and ease.

In scientific quests, it plays a role,
A catalyst for discoveries, a key to unfold.
From thyroid to radiology's might,
Iodine guides, a beacon of light.

Its enigmatic nature, a puzzle unsolved,
Mysteries within its depths, yet to be evolved.
A tapestry woven with Iodine's thread,
Unraveling wonders, awakening dread.

Oh, Iodine, you captivate, you inspire,
A substance of marvel, never to tire.
Unlocking the secrets of nature's design,
In your presence, wonders align.

So let us marvel at Iodine's might,
A chemical element shining bright.
From oceans to forests, mysteries unfold,
Iodine, a story waiting to be told.

FIVE

FOREVER WE CHASE

In the depths of the sea, where mysteries lie,
A shimmering element catches the eye.
Iodine, a jewel in nature's embrace,
Unveiling secrets with its ethereal grace.

Within the waves, where creatures reside,
Iodine dances, a cosmic guide.
Its atomic symphony, an enchanting song,
Weaving tales of wonder, where science belongs.

A master chemist, this element divine,
With prowess unmatched, it dares to define.
In the human body, it plays a vital role,
Regulating thyroids, harmonizing the soul.

From ancient times, alchemists sought its might,
In potions and spells, they reveled in its light.

A catalyst for change, a mystical key,
Iodine unlocks the wonders of chemistry.
    Through scientific endeavors, it has been revealed,
Iodine's power, its virtues unsealed.
From discoveries profound to breakthroughs untold,
Iodine's legacy, a story of gold.
    So let us marvel at this element rare,
Its beauty and power, beyond compare.
For in the realm of science and nature's embrace,
Iodine's secrets, forever we chase.

SIX

WONDERS OF LIFE

In nature's realm, where secrets lie,
A mystic hue beneath the sky.
Iodine, an element profound,
With power and beauty, so renowned.

    Within the depths of ocean's deep,
An iodine dance, a secret keep.
For in the waves, it does reside,
A treasure sought by curious tide.

    In forests green, where life abounds,
Iodine whispers without a sound.
It weaves its magic through the trees,
A symphony of balance it decrees.

    Within our bodies, it finds a home,
A vital spark, a thyroid's throne.

From metabolism to growth and more,
Iodine's touch, we can't ignore.

In labs of science, it holds the key,
Unlocking wonders for all to see.
From medicine's triumphs to cures untold,
Iodine's mysteries, forever unfold.

Enigmatic element, both bold and rare,
With sparks of brilliance, beyond compare.
Through fiery experiments, it shines so bright,
A beacon of knowledge, in scientific light.

Oh, iodine, a jewel in nature's crown,
A marvel of chemistry, so renowned.
With each discovery, we're left in awe,
In your essence, the wonders of life we draw.

SEVEN

TIME AND SPACE

In the realm of elements, a wondrous tale unfolds,
Of a noble substance, shimmering gold.
Iodine, a catalyst of growth and life,
With healing powers, it banishes strife.

Within our bodies, it quietly resides,
A guardian of balance, where health resides.
Thyroid's faithful companion, it takes its stand,
Regulating the rhythm of the body's grand.

From the depths of the sea, it emerges bright,
A gift from nature, a mesmerizing sight.
Bound in molecules, it weaves its spell,
Unlocking wonders, where science dwells.

A symbol of mystery, its secrets untold,
Iodine's presence, a story yet unfold.
Through centuries, it has captivated minds,
Unleashing truths, where innovation finds.

From art to medicine, its touch is felt,
A source of inspiration, where wonders are dealt.
Iodine, oh enigmatic element,
In your presence, the world finds content.

So let us marvel at your powers unseen,
And honor the legacy that you convene.
Iodine, the key to nature's embrace,
Unveiling the mysteries of time and space.

EIGHT

MAGIC

In shadows deep, where secrets hide,
A gleaming jewel, the iodine's pride.
With atomic might, it dances free,
Unraveling wonders, for all to see.

A catalyst of life, it breathes in the air,
Embracing cells, with tender care.
Thyroid's guardian, it whispers truth,
Balancing rhythms, in the body's booth.

Through alchemist's eyes, its tale unfolds,
A revelation of science, yet untold.
From Curie's hands, it sparked a fire,
Unveiling mysteries, that never tire.

In darkened labs, where scholars roam,
Iodine's dance, a heavenly poem.

It paints the world, with vibrant hues,
Mysterious, enchanting, an alchemic muse.
    Oh, iodine, in your atomic grace,
You leave a mark, on time and space.
A symbol of beauty, a beacon of might,
Unveiling the wonders, hidden from sight.
    From sea to sky, your presence shines,
A treasure sought, by curious minds.
In nature's embrace, you find your place,
A testament to wonders, science can trace.
    So let us marvel, at iodine's might,
A gift from the cosmos, shining so bright.
In every breath, and every thought,
Its magic weaves, a tapestry unsought.

NINE

COMPANION TO SCIENCE

In the depths of our being, a trace element resides,
An enigmatic force, where secrets do hide.
Iodine, the element, with powers untold,
A tale of wonder and mysteries unfold.

Within our bodies, it plays a vital role,
Regulating thyroids, keeping us whole.
Metabolism's guardian, it keeps us in tune,
A chemical conductor, orchestrating our own.

In labs and experiments, its significance is clear,
From early discoveries, its wonders appear.
A catalyst for change, unlocking new doors,
Revealing the marvels, that science explores.

Alchemy's dreams, once sought its embrace,
Transforming the ordinary, in nature's grace.

From colorless to vibrant, it dances with fire,
Transmuting the mundane, awakening desire.

In nature's domain, it lingers and weaves,
In seaweed and soil, where its essence perceives.
A whisper in the breeze, a shimmer in the sea,
Iodine's spirit, forever wild and free.

Oh, Iodine, captivator of minds,
Inspiring poets, their words it unwinds.
From deep within, its beauty unfurled,
A key to the wonders of this earthly world.

So let us celebrate this element divine,
For in its presence, we find the sublime.
Iodine, the enigma, forever to be,
A companion to science, a muse for me.

TEN

SWAY

In the realm of elements, a star does shine,
Iodine, a marvel, both rare and fine.
With atomic number fifty-three it stands,
A symbol of transformation in Nature's hands.

From the depths of the sea to the heights of the sky,
Iodine pervades, captivating the eye.
Its hue, a deep violet, a mystical hue,
Inspiring awe and wonder, as it comes into view.

In labs and in vials, its power is seen,
As it dances and swirls, a vibrant marine.
With its ability to transmute and create,
Iodine weaves its magic, altering fate.

In the poet's heart, it stirs up desire,
Igniting the flames, setting souls on fire.
Its presence in nature, a gift to behold,
An elixir of passion, both fierce and bold.

Oh, Iodine, enchantress of the elements,
A muse to poets, their words it augments.
With your atomic dance, you captivate and inspire,
A testament to the beauty that lies in the fire.

So let us celebrate this element of might,
Iodine, a beacon in the darkest night.
In its transformative embrace, we find our way,
To a world where wonder and magic hold sway.

ELEVEN

FOREVER IN OUR HEART

In the depths of the endless sea,
There lies a secret, hidden key.
A gleaming element, dark and rare,
Iodine, beyond compare.

Mysterious and enigmatic,
A symbol of the enigmatic,
It dances with the atoms, unseen,
Unleashing wonders yet unseen.

A savior to the human form,
Essential for life to be born,
Thyroid, thy master of control,
With iodine, it takes its toll.

Through ancient forests, it does roam,
In seaweed's embrace, it finds its home.

A trace of iodine, pure and bright,
A source of power, burning bright.
   From the laboratory's hallowed halls,
To scientific minds, it calls.
Discoveries made, knowledge gained,
With iodine, the world's explained.
   In potions and elixirs, it resides,
A potion of passion it provides.
Igniting fires within the soul,
Iodine, the flame, makes us whole.
   Oh, iodine, the element divine,
A symbol of the grand design.
Unlocking secrets, unveiling truth,
Iodine, the catalyst of youth.
   In every breath, in every cell,
Iodine's magic, it does dwell.
A testament to nature's art,
Iodine, forever in our heart.

TWELVE

SYMBOL OF POSSIBILITIES

In the realm of elements, Iodine shines bright,
A mesmerizing hue, a captivating sight.
With atomic number fifty-three, it declares its might,
An element so intriguing, a beacon of light.

From the depths of the sea, it rises with grace,
An elemental dance, a mystical embrace.
Its violet vapors swirl, leaving trails in space,
Unveiling secrets, unraveling the chase.

In the laboratory's realm, Iodine takes its stand,
A catalyst of change, a scientist's command.
Revealing unseen wonders, with a steady hand,
It unlocks the mysteries, like grains of sand.

Through art's vibrant canvas, Iodine finds its way,
A pigment of passion, igniting colors at play.

From sepia to azure, its hues gently sway,
Creating masterpieces, where emotions hold sway.

In the human body's tapestry, Iodine weaves,
A vital thread of life, where health perceives.
Regulating thyroids, as energy retrieves,
A guardian of balance, where wellness achieves.

Oh, Iodine, you captivate minds and souls,
With your enigmatic beauty, the tale unfolds.
From science to art, your influence extols,
A symbol of possibilities, as life unfolds.

THIRTEEN

SWEETEST WINE

In shadows deep, a mystic hue,
A potion brewed, both old and new.
A metal rare, a gleaming gem,
Iodine, the poet's requiem.

With power vast, it sparks the mind,
Unlocks the secrets, undefined.
A catalyst of thoughts profound,
In silent whispers, it is found.

Within the lab, its secrets lie,
Revealing truths, as time goes by.
From molecules to grand designs,
Iodine, the alchemist's signs.

In nature's realm, it paints the scene,
A touch of magic, subtle and serene.

The ocean's waves, a shade of blue,
Reflecting Iodine's eternal hue.

Inspiring artists, stroke by stroke,
Unleashing visions, as dreams evoke.
From canvas bare to masterpiece,
Iodine, the muse's release.

Within our bodies, it resides,
A vital spark, where life abides.
In thyroid's realm, a noble reign,
Iodine, the healer's domain.

Oh, Iodine, a force untamed,
In every realm, you leave your name.
A symbol of wonder, mystery untold,
A beauty rare, forever bold.

So let us marvel, in awe we stand,
At Iodine's touch, across the land.
A beacon bright, a spark divine,
Iodine, the poet's sweetest wine.

FOURTEEN

BEACON OF CREATIVITY

In the depths of the sea, a secret resides,
A shimmering element with transformative tides.
Iodine, the alchemist's delight,
Unleashing magic with its atomic might.

From ancient potions to elixirs of old,
Its essence weaves stories, untold and bold.
A catalyst for change, it sparks the flame,
Creating new worlds with its atomic name.

In laboratories, scientists toil,
Harnessing its power, unraveling the coil.
Unlocking mysteries, it guides their way,
Revealing truths that were once in disarray.

Within our bodies, it plays a role,
Regulating thyroids, a vital control.

Balancing the rhythms, harmonizing the beat,
Iodine, the conductor of life's sweet retreat.

In the realm of art, it takes the stage,
Inspiring creators, igniting a rage.
Brushstrokes of azure, a canvas alive,
Iodine's presence, a masterpiece to strive.

Oh, Iodine, a chemical dance,
A symphony of colors, a cosmic expanse.
In nature's embrace, you shimmer and glow,
A testament to the wonders you bestow.

So let us celebrate, this element divine,
Iodine, the catalyst, the muse, the sign.
Forever transforming, in science and art,
A beacon of creativity, forever in our heart.

FIFTEEN

ALWAYS BE THERE

In the depths of the ocean, where secrets reside,
There lies a shimmering element, a hue of the tide.
Iodine, the magician of molecules, they say,
With powers of transformation, it holds sway.

A catalyst of change, it dances with delight,
From solid to gas, from darkness to light.
It weaves its magic in the laboratory's embrace,
Unveiling mysteries, leaving none to efface.

In the realm of science, its wonders unfold,
As discoveries are made, its story is told.
From medicine to art, it leaves its mark,
A creative force, shining in the dark.

Within the human body, it plays a crucial role,
Regulating thyroids, keeping us whole.

A guardian of health, a balance it maintains,
Intricate as a melody, flowing in our veins.
    Iodine, a symbol of wonder and mystery,
A muse to poets, an enigma of history.
With every transformation, it captivates the soul,
Unleashing its magic, making us whole.
    So let us celebrate this element divine,
For its beauty and power will forever shine.
In the depths of the ocean, in the laboratory's lair,
Iodine, the alchemist, will always be there.

SIXTEEN

SILENT FORCE

In the depths of the ocean, where secrets abound,
A mystical element, Iodine is found.
With its shimmering hue, like the moon's gentle glow,
It enchants and captivates, a radiant show.

 A catalyst of change, it weaves its magic spell,
Transforming the ordinary, with stories to tell.
From dull to vibrant, from plain to sublime,
Iodine dances, an alchemist in time.

 In the artist's hand, it becomes a brushstroke bold,
Creating masterpieces, tales yet untold.
From canvas to sculpture, its presence imbued,
Iodine breathes life, in every shade and hue.

 In the lab, it's a scientist's guiding star,
Unraveling mysteries, taking us far.
From potions and mixtures, to breakthroughs pro-

found,
Iodine's knowledge, forever renowned.

   In the human body, it keeps balance intact,
A guardian of health, it knows no contract.
From thyroid to cells, it's an essential part,
Iodine's presence, a beating heart.

   Mysterious and enchanting, a symbol of wonder,
Iodine's beauty, we'll forever ponder.
In art and in science, its power is clear,
A silent force, forever held dear.

SEVENTEEN

ENCHANTING THE WORLD

In the realm of the periodic table's grace,
A radiant element holds its rightful place.
Iodine, the muse of mystique and might,
Unveils secrets in a celestial light.

A magician of atoms, it weaves its spell,
With a touch of iodine, wonders dwell.
Its symphony of electrons dances free,
Creating a cosmic symphony, for all to see.

In laboratories, scientists find solace,
As iodine reveals secrets, with its embrace.
It unlocks the mysteries of compounds unknown,
A catalyst for knowledge, forever grown.

Artists, too, find solace in its hue,
As iodine paints their canvases anew.

A splash of violet, a stroke of blue,
Unleashing creativity, bold and true.
   Within the human frame, it finds its home,
A guardian of health, it freely roams.
Balancing thyroids, it whispers its song,
Regulating harmony, all day long.
   In the depths of the ocean, it finds its peace,
Where seaweed sways, and creatures cease.
A symbol of life, in the depths below,
Iodine's presence, a constant glow.
   From the lab to the canvas, the body to the sea,
Iodine's influence, forever it will be.
A force of wonder, a catalyst divine,
Enchanting the world, through space and time.

EIGHTEEN

ITS PRESENCE

In shadows deep, a shimmering hue,
A mystic element, pure and true.
Iodine, the giver of light,
Igniting visions, both day and night.

An artist's brush, with colors bold,
Dips into Iodine, a story unfolds.
Canvas sings with vibrant hues,
Iodine's touch, creativity ensues.

In science's realm, a curious dance,
Iodine reveals, with every advance.
From lab to lab, its secrets unfold,
A catalyst of knowledge, yet untold.

Within the body, a vital role,
Iodine's presence, it takes control.
Thyroid's keeper, it regulates,
Health and balance, it orchestrates.

Yet deeper still, a wonder untamed,
Mysteries of Iodine, forever unnamed.
A symbol of magic, a token of awe,
It leaves its mark, in every draw.

Oh, Iodine, a jewel of the earth,
A catalyst of wonder, of infinite worth.
In art and science, you weave your spell,
A testament to the stories you tell.

So let us cherish, this element divine,
Iodine's essence, forever shall shine.
Inspiring awe, with each passing day,
In its presence, we shall forever sway.

NINETEEN

IODINE, THE ELEMENT

In the depths of the ocean's embrace,
Where secrets lie in the salted space,
A shimmering hue, a mystical trace,
Iodine, the element of grace.

A catalyst of life, it dances unseen,
Within our bodies, a vital routine.
Thyroid's keeper, it regulates the flow,
Metabolism's conductor, a rhythm to bestow.

But beyond the realm of biology's might,
Iodine weaves its spell, a creator of light.
In the artist's hand, a stroke of blue,
A canvas transformed, a masterpiece anew.

From tinted photographs of days gone by,
To the stained glass windows that touch the sky,

Iodine's essence, a pigment divine,
A testament to its power, forever shall shine.

    Yet, in the realm of science's quest,
Iodine remains an enigmatic guest.
Its properties, a puzzle to unravel,
A mystery that scientists forever grapple.

    Oh Iodine, a paradox of intrigue,
A whisper of the universe's fatigue.
From healing to wonder, you hold the key,
An element of magic, forever to be.

    So let us marvel at your boundless might,
In awe of your presence, day and night.
For in your essence, we find our way,
Iodine, the element that forever holds sway.

TWENTY

HUSH

In the realm of elements, a jewel divine,
Iodine, the enchantress, doth brightly shine.
With atomic number fifty-three, she stands,
A mistress of mystery in her hands.

In the depths of the ocean, her essence lies,
As waves crash and tumble 'neath cloudy skies.
Her iodide ions dance with the sea,
A symphony of life, a melody set free.

In the human body, she claims her reign,
A guardian of health, a cure to pain.
Thyroid's faithful ally, she plays her part,
Regulating our metabolism with her art.

Artists and scientists, they both adore,
This element, which sparks their souls to soar.

From Renaissance painters to modern minds,
Iodine's allure, forever it binds.

   In laboratory flasks, her colors unfold,
From vibrant violet to shimmering gold.
Her chemical reactions, a dazzling sight,
Igniting the imagination, filling hearts with light.

   Oh Iodine, thy beauty knows no bound,
In nature's realm, a treasure to be found.
From the depths of the ocean to the artist's brush,
Your presence, forever, an enchanting hush.

TWENTY-ONE

WONDERS OF IODINE

In depths where ocean's secrets hide,
A realm of wonder, deep and wide.
Iodine, an element profound,
A guardian of health, we have found.

Within our bodies, you reside,
Regulating, a constant guide.
Thyroid's keeper, you maintain,
Balance of life, a sacred reign.

In art and science, you inspire,
A catalyst, a creative fire.
With hues of blue, like ocean's hue,
You bring forth visions, ever true.

Mysterious element, enigmatic grace,
Unveiling secrets, in every trace.

From ancient times, you've been revered,
A symbol of wisdom, forever endeared.
    Oh Iodine, in your beauty we see,
The power of nature's alchemy.
A potent force, forever blessed,
In science and art, you manifest.
    So let us celebrate thy name,
In awe of your eternal flame.
Through art and science, we shall explore,
The wonders of Iodine, forevermore.

TWENTY-TWO

CHEMICAL MARVEL

In the depths of art, where colors dance,
A hue emerges, a captivating trance.
Iodine, the alchemist's delight,
Ignites the canvas, a vibrant light.

From sea to sky, its essence unfurls,
A mystery held by nature's pearls.
In the depths of the ocean, where creatures roam,
Iodine whispers secrets, the depths it calls home.

A catalyst of health, a guardian of thy,
It regulates the gland, where secrets lie.
Thyroid's keeper, a conductor of grace,
Iodine ensures balance in life's intricate chase.

In the laboratory, where science thrives,
Iodine reveals its atomic archives.

With a gleaming symbol, I, it stands,
A chemist's treasure, between their hands.
   Its electrons dance, in valence they play,
An enigma of elements, in their array.
A halogen, a wonder, a sublime embrace,
Iodine's allure, no scientist can erase.
   From the brushstrokes of art to the depths of the sea,
Iodine's presence, a marvel to see.
A catalyst for wonder, an elixir of awe,
This chemical marvel, forever we adore.

TWENTY-THREE

MUSE SUPREME

In the depths of the ocean's embrace,
Where secrets linger, in hidden grace,
Iodine dwells, mysterious and rare,
A catalyst for wonder, a cosmic affair.

A guardian of balance, it holds the key,
Regulating thyroids, in harmony,
With gentle touch, it calms the storm,
A healer of bodies, a balm for the worn.

In the realm of science, it takes flight,
Unraveling mysteries, unveiling light,
Its atomic dance, a mesmerizing sight,
Igniting curiosity, like stars in the night.

But Iodine's allure extends beyond the lab,
It dances with artists, a muse to grab,
With vibrant hues, it paints the canvas,
Stirring emotions, like love's first kiss.

In the realms of creativity, it weaves its spell,
Unleashing imagination, where stories dwell,
From poets' pens to sculptors' hands,
Iodine's essence, forever expands.

So let us marvel at this enigmatic element,
A symphony of beauty, a poetic testament,
For in its depths, both seen and unseen,
Iodine reigns, a muse supreme.

TWENTY-FOUR

ENTHRALL

In the depths of the ocean, a secret lies,
A radiant jewel, that gleams and defies,
Iodine, the element of mystery untold,
A tale of wonder, waiting to unfold.

A catalyst for life, in the thyroid it dwells,
Regulating the rhythm, the stories it tells,
Balancing the body, with grace and finesse,
Iodine, the conductor, in perfect caress.

In art and science, it leaves its mark,
A hue of violet, igniting the dark,
From photographs to stained glass windows,
Iodine's presence, a masterpiece bestows.

Oh, enigmatic element, elusive and rare,
Captivating hearts, with a mystical flair,
With electrons dancing, in a graceful waltz,
Iodine, the muse, in the poet's exalts.

Curiosity sparked, by its atomic might,
Unveiling the secrets, hidden from sight,
Inspiring the mind, with a creative surge,
Iodine, the catalyst, that makes us urge.

In the laboratory, where discoveries bloom,
Iodine's essence, a scientist's room,
Unraveling the universe, molecule by molecule,
Iodine, the guide, in the quest to excel.

Oh, Iodine, enchanting and pure,
A symbol of grace, that will endure,
With elegance and power, you captivate all,
Iodine, the element, in beauty you enthrall.

TWENTY-FIVE

IT SHALL BE

In the realm where science intertwines with art,
A substance emerges, playing its part.
Iodine, a chemical with allure,
Regulating metabolism, its nature pure.

A catalyst for the body's grand symphony,
It orchestrates thyroxine, a metabolic decree.
From the thyroid's throne, it reigns supreme,
Balancing the dance of life's endless dream.

Yet beyond the realm of the scientific few,
Iodine captures hearts, imagination it imbues.
In the artist's palette, a radiant hue,
Inspiring strokes, a masterpiece anew.

Scientists marvel at its atomic grace,
Curiosity ignited by its mysterious embrace.

A symbol of wonder, it sparks the mind,
Seeking truths that are difficult to find.

    Oh, Iodine, muse of science and art,
Whispering secrets from the depths of your heart.
A symbol of grace, you captivate all,
Igniting creativity, like a siren's call.

    So let us celebrate this element divine,
With awe and wonder, let our spirits align.
For in Iodine's presence, we find solace and glee,
A symbol of beauty, forever it shall be.

TWENTY-SIX

WONDER AND GRACE

In the depths of the ocean blue,
A hidden gem emerges to view.
Iodine, a mystic element of grace,
Whispers secrets of the ancient space.

Its shimmering hue, a cosmic charm,
Ignites the minds, sets hearts to warm.
With electrons dancing, it creates a spell,
A symphony of science, where wonders dwell.

A muse for artists, it paints the world,
With shades of purple, its flag unfurled.
From stained glass windows to vibrant dyes,
Iodine's touch, a feast for the eyes.

In laboratories, it sparks the fire,
A catalyst for discovery, taking us higher.

From medicine to photography's art,
Iodine's essence, a masterpiece, a part.

It bonds with atoms, a dance of delight,
Unveiling secrets, like stars in the night.
Its atomic number, a number divine,
Iodine, a symbol, a treasure to find.

So let us celebrate this element rare,
Its beauty and mystery, beyond compare.
In science and art, it takes its place,
Iodine, a symbol of wonder and grace.

TWENTY-SEVEN

WILL STAND

In the realm of elements, let me sing,
Of Iodine, the catalyst of dreams.
With atomic number fifty-three,
It unveils beauty, for all to see.

In laboratories, its hues align,
A vibrant violet, a shade divine.
A catalyst for change, it plays its part,
Igniting reactions, a fiery art.

But beyond the walls of science's might,
Iodine dances in artistic light.
A muse for poets, a stroke of grace,
It weaves its elegance with every trace.

In the depths of oceans, its presence found,
A symbol of life, profound and profound.

From kelp to sea creatures, its embrace,
A testament to nature's wondrous grace.
    Yet Iodine's allure extends much more,
It captivates minds, rich with explore.
For scientists and artists, it's the same,
A wellspring of inspiration's flame.
    So let us celebrate this element rare,
With its mysteries and wonders to share.
From the laboratory to the artist's hand,
Iodine, the muse, forever will stand.

TWENTY-EIGHT

SCIENCE AND DREAM

In the depths of science's realm,
A luminary, Iodine, takes the helm.
A catalyst for creativity's spark,
Unraveling mysteries, igniting the dark.

With elegance, it dances in the lab,
A potent elixir, a scientist's fab.
Its vibrant hues, a painter's delight,
Brushing art with a captivating light.

An enigma it is, this element pure,
A symbol of wonder, forever endure.
A muse to poets, their verses it inspires,
Ink flowing freely, as passion transpires.

From the iodine clock, time unfurls,
A symphony of reactions, a dance of swirls.
In chemistry's embrace, it finds its place,
A symbol of grace, in every trace.

Nature's ally, it adorns the sea,
Where waves crash, with a salty plea.
A trace in the air, a scent so pure,
A reminder of nature's allure.

Oh, iodine, a gem in the periodic table,
Your essence, a wonder that none are able,
To resist your charm, your magical gleam,
Forever captivated, in science and dream.

TWENTY-NINE

CREATIVITY'S REALM

In the depths of the ocean blue,
Where mystery and wonder grew,
There lies a secret, shining bright,
A jewel of nature's endless light.
 Iodine, a radiant hue,
A poet's muse, forever true,
Inspiring words, like waves that crash,
A symphony of thoughts, a vibrant splash.
 Its atomic dance, a graceful glide,
Igniting curiosity far and wide,
A catalyst for scientific minds,
Unveiling truths that nature binds.
 A painter's brush, a poet's pen,
Capturing Iodine's essence, again,

In hues of purple, deep and rare,
A palette of beauty beyond compare.
   Oceans hold its precious might,
A symbol of grace, a precious sight,
In waves that crash upon the shore,
Iodine whispers, forevermore.
   A symbol of wonder, both fierce and kind,
In chemistry's realm, a jewel we find,
An element of elegance, it gleams,
Iodine, the muse of dreams.
   So let us marvel, let us explore,
In Iodine's embrace, forevermore,
For in its depths, we find our way,
To creativity's realm, where wonders sway.

THIRTY

IODINE, THE ELEMENT THAT IGNITES

In the realm of art, science, and nature's grace,
Lies a chemical element, Iodine, with a mystic trace.
A dancer on the periodic table's stage,
Its allure captivates both hearts and minds engage.

    In laboratories, it's a catalyst's delight,
Aiding reactions with its atomic might.
From organic compounds to pharmaceutical feats,
Iodine's presence in chemistry never retreats.

    But beyond the confines of scientific lore,
It leaves its mark on artists who explore.
A hue of blue, a touch of violet's fire,
Iodine inspires painters' creative desire.

From swirling skies to ocean's deep embrace,
Its pigment lends beauty to each painted space.
For poets, it's a symbol of passion's flame,
Igniting verses with its lyrical name.

In the natural realm, it's a gift so divine,
Found in seaweed, in the depths of the brine.
A trace element essential for life's grand design,
Iodine's presence echoes nature's rhyme.

Oh, Iodine, you hold secrets untold,
Invisible to the naked eye, yet so bold.
A symbol of wonder, a mystery profound,
Your beauty and power forever astound.

So let us celebrate this element rare,
For its presence in our world, we must declare.
From laboratories to canvas and sea,
Iodine's allure shall forever be.

In art, in science, in nature's embrace,
Iodine's essence leaves an indelible trace.
A testament to its enduring might,
Iodine, the element that ignites.

THIRTY-ONE

LEAVE US IN AWE

In the depths of the ocean, where secrets reside,
Where waves crash and currents collide,
There lies a shimmering element, so rare,
With a mystique that captures hearts, I swear.

Iodine, oh Iodine, how you mesmerize,
With your deep blue hue and enchanting guise,
A symbol of wonder, a chemical delight,
You ignite our imagination, day and night.

In the laboratory, you dance with grace,
Revealing reactions, a chemical embrace,
Your atomic number, fifty-three,
Unveils the magic you bring to chemistry.

But beyond the lab, your beauty transcends,
In art and nature, a connection that never ends,
In vibrant sunsets, you paint the sky,
A kaleidoscope of colors, oh my!

You're a muse to poets, a melody to song,
A spark of inspiration, forever strong,
Your presence whispers of tales untold,
Of ancient wisdom and secrets unfold.

So, Iodine, we celebrate your splendor,
Your elegance, your power, so tender,
In science, art, and nature's embrace,
You leave us in awe, your allure we chase.

THIRTY-TWO

SPARK

In the depths of the ocean, where mysteries lie,
A wondrous element catches the eye.
Iodine, the muse of the curious soul,
A symbol of wonder, with secrets untold.
    A catalyst for chemistry's dance,
Igniting reactions with each little chance.
From the lab to the world, its influence profound,
In medicines and dyes, its presence is found.
    A hue of the night, a delicate blue,
Iodine's essence, it enchants and renews.
A siren's song in the artist's brush,
Creating masterpieces with every soft touch.
    Nature's own magic, it weaves and it spins,
A gift to the earth, where life's journey begins.

From the sea to the land, its reach does extend,
A vital connection, a bond without end.
    Oh, Iodine, you inspire awe and delight,
With your shimmering glow, so pure and so bright.
From the tiniest atom to the vast cosmic sky,
You captivate hearts with your endless supply.
    So let us embrace this element divine,
And marvel at its beauty, so rare and so fine.
For in Iodine's presence, we find our own grace,
A spark of creativity, a world to embrace.

THIRTY-THREE

UNFOLD

In the depths of the ocean, where secrets lie,
There dwells a shimmering element, Iodine by name.
A mystery it carries, a tale untold,
A beauty that only the sea can unfold.

Its hue, a deep indigo, like twilight's embrace,
Iodine dances with the waves, a celestial grace.
Its essence, ethereal, it whispers in the breeze,
Awakening senses, igniting souls with ease.

In chemistry's embrace, it wields its might,
A catalyst of change, a symphony of light.
A dance of electrons, a ballet of bonds,
Iodine weaves its magic, chemistry responds.

Artists are enchanted, their canvases ablaze,
As Iodine's touch transforms their creative haze.

Colors burst forth, vibrant and wild,
Iodine's inspiration, an artistic child.
    In the laboratory, scientists delve,
Unraveling Iodine's secrets, they strive to excel.
From medicines to dyes, it lends its hand,
Iodine, the alchemist's dreamland.
    Oh, Iodine, enigma divine,
In your presence, hearts and minds intertwine.
A symbol of wonder, elegance untold,
Iodine, your allure will forever unfold.

THIRTY-FOUR

ESSENCE ENDURES

In the depths of the ocean, hidden and deep,
Iodine, the secret, its mysteries we seek.
A shimmering hue, like the midnight sky,
It captivates hearts, never asking why.

A catalyst of art, it paints the canvas bright,
Staining our minds with its ethereal light.
From the depths of our souls, creativity flows,
Iodine's inspiration, a masterpiece it bestows.

In the darkest of nights, it guides the way,
A lighthouse of knowledge, leading astray.
Chemical reactions, it sparks and ignites,
Unlocking the secrets, unveiling the sights.

The sea's salty embrace, where it finds its home,
Iodine dances and roams, never to be alone.
It sings with the waves, harmonizing with glee,
A symphony of elements, an ode to the sea.

A healer it becomes, when illness takes hold,
Iodine, the remedy, its power never untold.
Restoring vitality, with its gentle touch,
A guardian of health, it loves us so much.

So let us celebrate, this element divine,
Iodine, the enigma, forever it will shine.
From art to science, nature to cure,
Its essence endures, timeless and pure.

THIRTY-FIVE

SHINING EVER BRIGHT

In the depths of the ocean, where the waves crash and roar,
Lies a secret in the water, a treasure to explore.
Iodine, the element of mystery and might,
A canvas for creativity, a beacon in the night.

In the artist's hand, a brush dipped in its hue,
Painting vivid landscapes, dreams brought to life anew.
A stroke of blue and purple, a touch of ethereal grace,
Iodine whispers inspiration, a muse in every trace.

In the laboratory, its power is revealed,
A catalyst of change, reactions unconcealed.
With precision and expertise, scientists unfold,
The secrets of Iodine, a story yet untold.

In medicine it triumphs, a healer in disguise,

Fighting silent battles, where hope never dies.
From antiseptic wonders to thyroid's gentle care,
Iodine, a savior, a remedy beyond compare.

Its allure is timeless, its presence ever strong,
Like a lighthouse in the darkness, guiding ships along.
From the depths of the ocean to the skies up above,
Iodine, a symbol of beauty, a testament of love.

So let us raise our voices, in praise and in awe,
For Iodine's enchantment, forever we shall draw.
A symphony of wonder, a dance of pure delight,
Iodine, the element, shining ever bright.

ABOUT THE AUTHOR

Walter the Educator is one of the pseudonyms for Walter Anderson. Formally educated in Chemistry, Business, and Education, he is an educator, an author, a diverse entrepreneur, and he is the son of a disabled war veteran. "Walter the Educator" shares his time between educating and creating. He holds interests and owns several creative projects that entertain, enlighten, enhance, and educate, hoping to inspire and motivate you.

Follow, find new works, and stay up to date
with Walter the Educator™
at WaltertheEducator.com

www.ingramcontent.com/pod-product-compliance
Lightning Source LLC
LaVergne TN
LVHW051959060526
838201LV00059B/3734